edZOOcation
Sharks

by Sara Karnoscak

Dedication:

For Ryder and Elijah.

–S.K.

For Jordan

–A.R.

Copyright © 2023 Wildlife Tree, LLC. All rights reserved.

Author: Sara Karnoscak

Designer: Allyson Randa

Editor: Tess Riley

Photo Credits:

AdobeStock.com

Pixabay.com

Pexels.com

Vecteezy.com

ISBN: 979-8-9859544-8-7

This book meets **Common Core** and **Next Generation Science Standards.**

Table of Contents

4	A Shark's Body
8	Shark Prints and Poop
10	Shark Species
14	Shark Pups
16	Shark Talk
17	Where Sharks Live
21	Sleeping and Eating
24	A Day in the Life of a Shark
26	The Food Web and Dangers
28	Shark Fun
30	Glossary
32	Silly Sharks

A Shark's Body

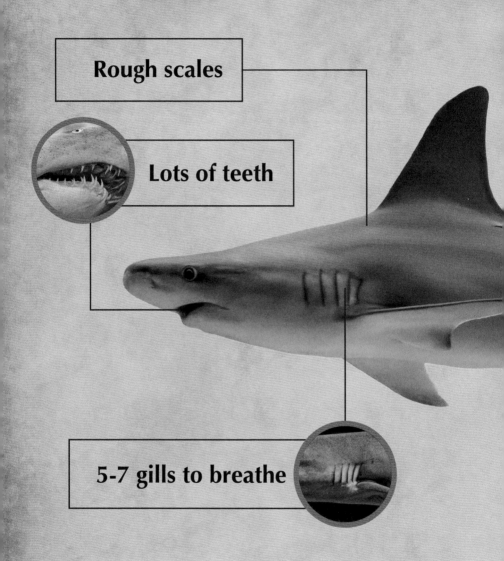

Rough scales

Lots of teeth

5-7 gills to breathe

A shark loses thousands of teeth in its life. They can grow back in a day.

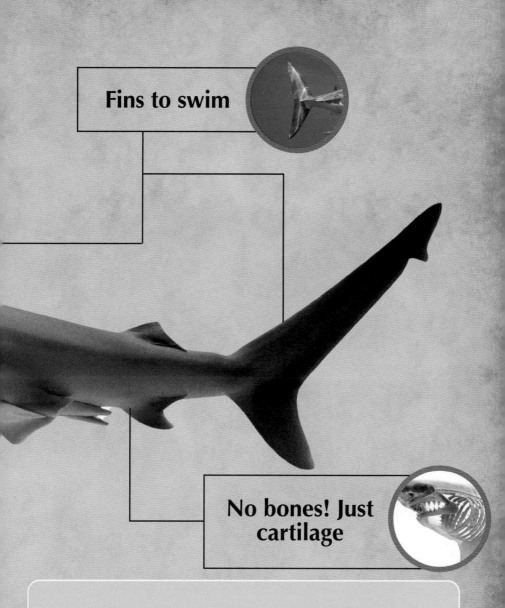

Fins to swim

No bones! Just cartilage

Cartilage: The bendy stuff in your nose and ears. Try bending your ears. Sharks can bend, too!

Shark Powers

Sharks have super senses.

Great Sight: Shark eyes have two parts. One to see in light. One to see in dark.

A Sixth Sense: Sharks can sense **electric fields**. They can use them as a map. They can use them to find **prey**.

Clean Teeth: Sharks have a coating on their teeth that's like toothpaste.

Electric Field: Force made by movement or something electric.

Prey: Animals that are hunted by other animals.

Sharks were around before the dinosaurs! How do you think sharks have survived for so long?

Shark Fashion

Not all sharks are gray. **Leopard sharks** have spots.

Tiger sharks have stripes.

Zebra sharks have stripes when they're young. But they have spots when they get older.

Pyjama sharks look like they are wearing striped pajamas.

Ew!

Shark poop is green! Smaller fish like to eat shark poop.

Some sharks can feed on the oil in their bodies. But the oil helps them float. So, when they feed on the oil, they sink more.

Big Shark, Little Shark

There are about 500 shark species.

Whale sharks are the biggest sharks. They only eat tiny **plankton**.

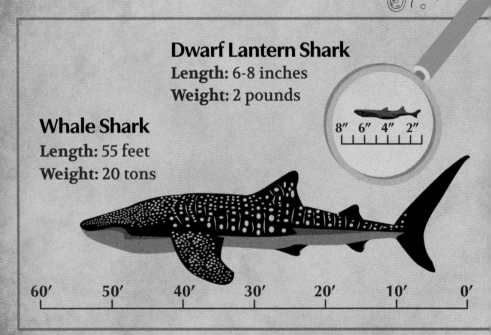

Dwarf Lantern Shark
Length: 6-8 inches
Weight: 2 pounds

Whale Shark
Length: 55 feet
Weight: 20 tons

Dwarf lantern sharks are the smallest sharks. They are as long as a hand.

Plankton: Tiny living things floating in the ocean.

Old Sharks

Megalodons used to be the biggest sharks. They lived in the time of the dinosaurs.

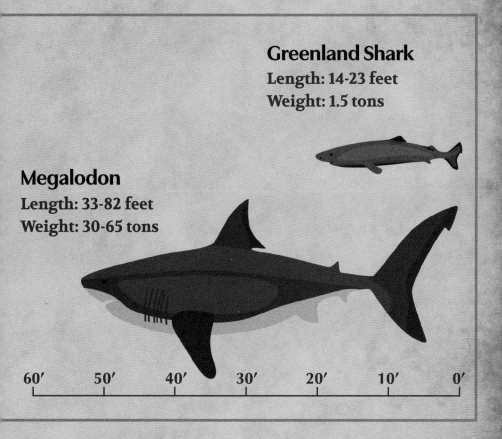

Greenland Shark
Length: 14-23 feet
Weight: 1.5 tons

Megalodon
Length: 33-82 feet
Weight: 30-65 tons

Greenland sharks are the longest-living sharks. They can live to be 500 years old.

Volcano Sharks?

Hammerhead sharks are the best swimmers. Their heads help them swim.

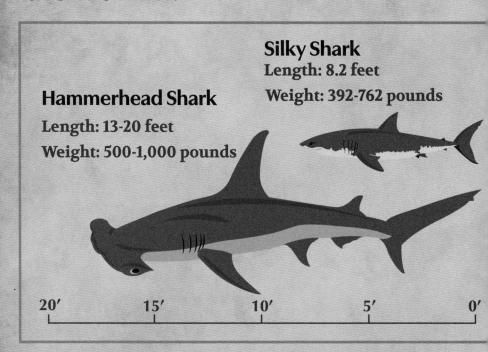

Silky Shark
Length: 8.2 feet
Weight: 392-762 pounds

Hammerhead Shark
Length: 13-20 feet
Weight: 500-1,000 pounds

20' 15' 10' 5' 0'

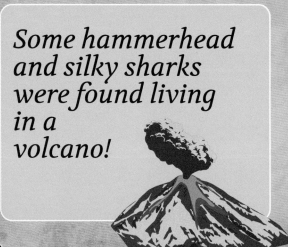

Some hammerhead and silky sharks were found living in a volcano!

Silky sharks are the smoothest sharks. Their scales don't feel rough.

Match the Size

How long is each shark? Draw a line with your finger.

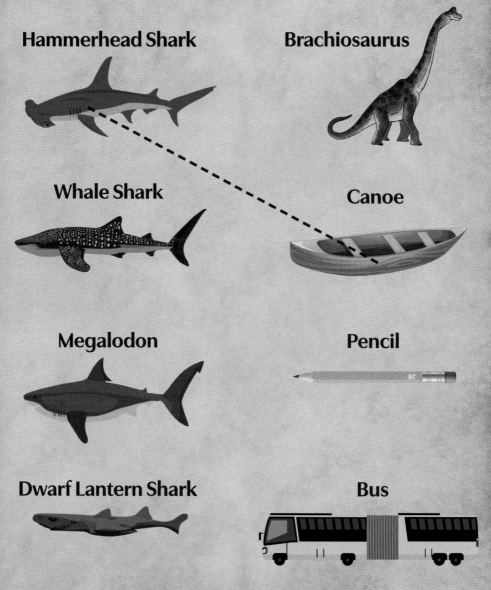

Hammerhead Shark

Brachiosaurus

Whale Shark

Canoe

Megalodon

Pencil

Dwarf Lantern Shark

Bus

Whale Shark & Bus, Megalodon & Brachiosaurus, Dwarf Lantern Shark & Pencil

Shark Pups

Some sharks lay eggs in a **mermaid's purse**.

Some sharks lay eggs in their bodies. The **pups** hatch and grow inside their mom. She lays more eggs for them to eat.

Pup: A baby shark.

Mermaid's purse: The case where a shark pup grows.

Pups are born with teeth. They go out on their own as soon as they are born.

How Do Sharks Talk?

Sharks can't make noise. They talk with their bodies.

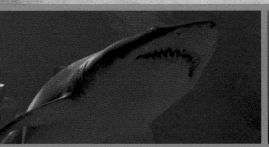

They open their mouths.

They fluff their gills.

They slam into each other.

They turn their bodies.

A Shark's Home

Some sharks live in one place.

Some sharks travel across the ocean.

Sharks Live All Over The World

They live in warm water.
They live under the Arctic ice.

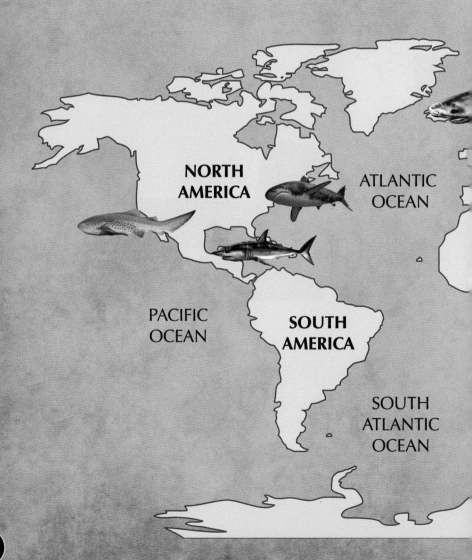

They live in shallow waters.
They live in the deep sea.

Word Jumble

Unscramble the letters to create the correct word. Write your new word on the line.

1. SIFN _____

2. ILLGS _____

3. ETTHE _____

4. ONEAC _____

5. SIFH _____

1. Fins 2. Gills 3. Teeth 4. Ocean 5. Fish

Sleep Swimming

Some sharks never stop moving. Even when they sleep! These sharks breathe by moving.

They may swim up and take a nap as they go down.

A Shark's Favorite Food

Sharks like to eat…

FISH

SEALS

BIRDS

RAYS

Sharks don't like to eat humans.

Sharks taste-test their food. Humans don't pass the taste-test.

A Day in the Life...

In the morning, the shark is hungry. He hasn't eaten in a few days.

His sixth sense tells him there is movement below.

# of a Great White

He spots a ray below him.

He uses his many teeth to catch the ray.

It tastes nice and oily.

Soon he will start to swim across the ocean. The oil will be good food on his journey.

The Shark Food Web

Sharks keep the ocean healthy. If sharks didn't hunt other animals, the ocean would be unhealthy.

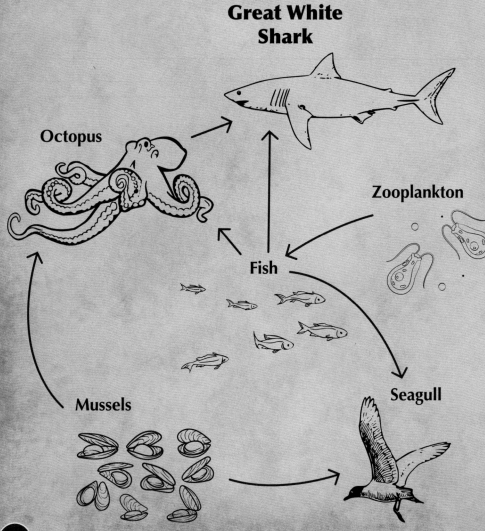

Dangers

The biggest danger to sharks is humans. Humans kill about 100 million sharks a year.

The odds of getting bitten by a shark? One in 11.5 million.

Can You Act Like a Shark?

Sharks use their bodies to talk. Use your body to talk.

Use your body to say, **"I'm curious."**
Now say, **"I'm scared."**
Now say, **"Stay away."**

Some sharks never stop moving. How long can you keep moving?

If you were a shark, what kind of shark would you be?

Glossary

Cartilage: The bendy stuff in your nose and ears.

Electric Field: Force made by movement or something electric.

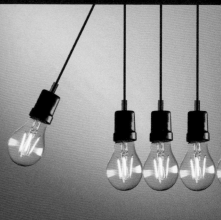

Mermaid's Purse: The case where a shark pup grows.

Plankton: Tiny living things floating in the ocean.

Prey: Animals that are hunted by other animals.

Pup: A baby shark.

Silly Sharks

The shark kept the ocean clean.
He didn't ever mean
To scare all the swimmers
They weren't his dinner
Their taste made his gills quite green.

What did the shark say when he ate the clownfish?

"This tastes funny!"

Why do sharks swim in salt water?

Because pepper water would make them sneeze!

Did you hear about the all-star shark athlete?

He led his team to the **chomp**ionship!

What's a shark's favorite breakfast food?

A jellyfish filled donut.

Why did the shark cross the reef?

To get to the other tide!

What's a shark's favorite card game?

Go fish.

Become a
Wildlife Guardian

Everyone can be a Wildlife Guardian by helping our planet. When we help our planet, we help the animals who share it with us.

Here are ways you can help sharks:

- Take care of the Earth. Recycle or re-use things when you can.

- Learn about the animals you love. The more you learn, the more you can help!

- Teach others about the animals you love. The more they learn, the more they can help, too!

- Don't eat shark meat.

- Join a beach or park cleanup. Even trash that's not on the beach can be carried there by rivers and waterways.

edZOOcation.com

edZOOcation™ Rea...
Inspiring Love for Animals Through ...

Early Reader (Pre-Reader) books use pictures and familiar ideas to teach simple scientific concepts to kids just *beginning to read with help*.
Repetition • Simple Text • Large Print

Reading Together (Level 1) books use pictures and familiar ideas to teach primary scientific concepts to kids *beginning to sound out words and recognize sight words*.
Limited Vocabulary • Sight Words • Simple Sentences

Supported Reader (Level 2) books introduce new ideas, supported by pictures, to teach more advanced scientific concepts to kids who are *beginning to read more independently*.
New Vocabulary • Longer Sentences • Abstract Topics

Independent Reader (Level 3) books introduce new ideas, supported by pictures, to teach more advanced scientific concepts to kids who are *fluent, independent readers*.
Advanced Text • Complex Sentences • Abstract Topics

edZOOcation.com

US $ 5.99
ISBN 979-8-9859544-8-7

Let's Play!

Judi Laman

Vocabulary

outside
seasons
sports
weather

Theme: Beyond My World
Word count: 368

Glenview, Illinois
Boston, Massachusetts
Chandler, Arizona
New York, New York

www.rubiconpublishing.com

Copyright © 2020 Rubicon Publishing Inc. Published by Rubicon Publishing Inc., developed in collaboration between Rubicon Publishing Inc. and Pearson Education, Inc., the exclusive United States distributor. All rights reserved. No part of this publication may be reproduced or transmitted in any form or by any means, electronic or mechanical, including photocopying, recording, taping, or any information storage and retrieval system, without the prior written permission of the copyright holder.

Associate Publisher: Amy Land
Executive Editor: Teresa Carleton
Managing Editor: Dawna Brochu
Editorial Assistant: Claire Christopher
Creative Director: Jennifer Drew
Lead Designer: Jason Mitchell
Designer: Jennifer Foreshew

Printed   in   Mexico
2   19
ISBN 978-1-4869-0559-1